Canyons in N. Africa:

5. Morocco
6. Algeria
7. Libya
8. Tunisia
9. Egypt
10. Sudan

Canyons in Asia:

11. India
12. China
13. Saudi Arabia
14. Iraq
15. Iran
16. Israel
17. Pakistan

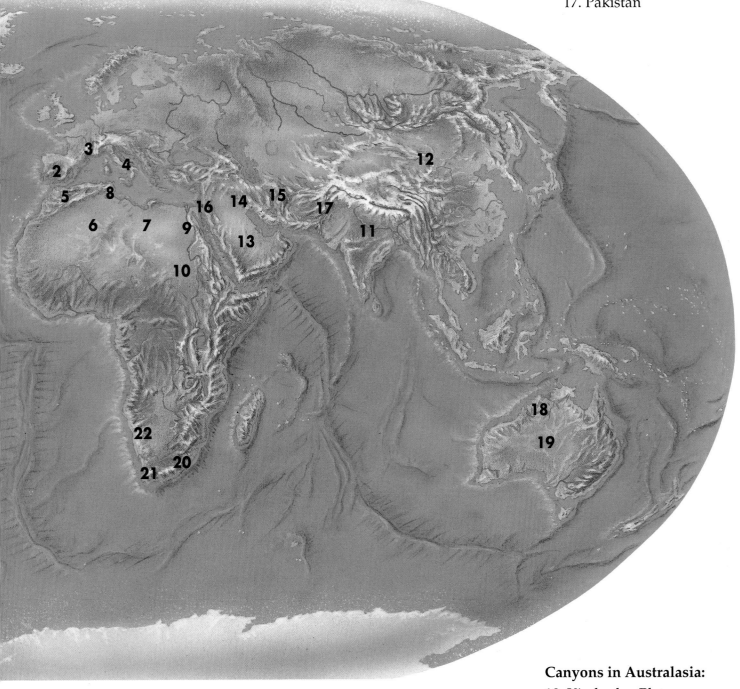

Canyons in Australasia:

18. Kimberley Plateau
19. Macdonnell Range

3

FACTS ABOUT CANYONLANDS

Canyons (also called wadis in North Africa) are gorge-like valleys in dry lands. The world's largest canyon is the Grand Canyon in Arizona, USA, and it is cut into a plateau nearly 8000 ft high. It is nearly 217 miles long and is over 5300 ft deep. The river Colorado, in forming the canyon, has cut through rocks that have existed for over 1 billion years of the Earth's history.

The world's longest natural arch is in Arches National Park, Utah, USA. Landscape arch is 291 feet long and about 100 feet above the ground. Rainbow bridge, Utah, is a natural bridge 278 ft wide. Another natural bridge in China has a span of 150 ft over a gorge 1000 ft deep.

Sudden storms cause flash flooding in canyons and people camping on canyon floors have been drowned because they have not been able to scramble to safety.

Grolier Educational Corporation
SHERMAN TURNPIKE, DANBURY, CONNECTICUT 06816

LAND  SHAPES
CANYON

Author
Brian Knapp, BSc, PhD
Art Director
Duncan McCrae, BSc
Editor
Rita Owen
Illustrator
David Hardy
Print consultants
Landmark Production Consultants Ltd
Printed and bound in Hong Kong
Designed and produced by
EARTHSCAPE EDITIONS

First published in the USA in 1993 by
GROLIER EDUCATIONAL CORPORATION,
Sherman Turnpike, Danbury, CT 06816

Library of Congress #92–072045

Cataloging information may be obtained
directly from Grolier Educational Corporation

Title ISBN   0–7172–7180–3

Set ISBN   0–7172–7176–5

Acknowledgements. The publishers would like
to thank the following: Leighton Park School,
Martin Morris, Redlands County Primary
School and Rob Lindgren of Redtail Aviation.

Picture credits. All photographs from the
Earthscape Editions photographic library.

Cover picture: Goosenecks State Park, Utah, USA.

In this book you will find some words that have been shown in **bold** type. There is a full explanation of each of these words on page 36.

On many pages you will find experiments that you might like to try for yourself. They have been put in a blue box like this.

In this book mi means miles and ft means feet.

These people appear on a number of pages to help you to know the size of some landshapes.

CONTENTS

Introduction

A canyon is a deep, steep-sided valley that forms in those parts of the world where there are long dry seasons. Rivers may keep running, supplied from distant mountains or even from water that slowly seeps through thick rocks, but on the surface all is barren and dry.

Rain falls on the canyonlands in short, torrential bursts, and as there is so little soil to soak up all this water, it soon rushes and tumbles over the surface, carrying any loose rocks with it.

Canyon rivers can be dry one minute and raging torrents the next. Rain and rock make a formidable mixture with immense power to wear away, or **erode**, the land. These floods are so powerful that large objects such as trucks and even houses have been carried away and smashed.

The world's biggest canyon, the Grand Canyon, is over one mile deep and thirteen miles wide, but there are many other canyons, big and small, around the world. In this book you can find out about the world's canyons, you can explore their depths and see how they were formed. You can also find out how canyons are just one part of a dramatic landscape of cliffs, arches, natural bridges and needle-shaped rocks. Just turn to a page to enjoy the wonder of canyonlands.

Take care in canyonlands ⚠

Canyonlands are some of the world's most exciting landshapes and you are sure to want to visit them. But never go down into a canyon without taking an adequate water supply, wearing the correct footwear and accompanied by an adult. Canyons can be dangerous places for the unwary and deaths have occurred because people have hiked canyons unprepared.

Chapter 1
How canyons form

What rivers do

Even though canyon country is dry for most of the year, much of the landscape is still shaped by rivers. In some places rivers flow all the time, carrying their load of pebbles and sand and using it to scour at the bed and banks; in other places rivers flow only for a few hours or days after a storm. In every case the scouring action is nature's way of cutting into the rock. This is the effect.

For more information on the work of rivers see the book River *in the Landshapes set.*

Rivers cut into their beds, making the **trench** that is the start of a canyon (see page 12).

The narrow neck between river curves may be cut through to form a natural bridge (see page 14).

The yellow color of the water tells you that the river is carrying enormous amounts of rock and soil. This means it must be cutting quickly into the landscape.

'White water' is a sign of power in a river. It occurs when water flows so quickly that it makes waves.
Here a pleasure raft is tossed about like a plaything.

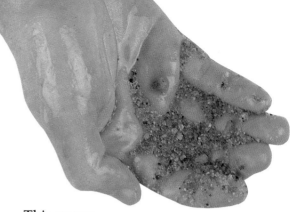

This coarse sand was collected from the bed of the river shown on this page.

As the river sweeps around the bend it cuts further into the cliff and more rock tumbles into the channel (see page 12).

As rivers cut into the **tableland** they start to carve it up into seperate 'islands' of rock. In time this will produce **mesas** and **buttes** (see pages 22 and 26).

See how rivers erode

Rocks near the banks and on the river bed are smoothed off (scoured) by the scraping action of pebbles and sand carried in the water.

All sizes of material can act as natural 'sandpaper'. It is not necessary for a river to carry large boulders in order to cut into its bed quickly, sand grains will also do a very efficient job.

You can see how this happens by rubbing sandpaper on a piece of rock. Notice that both sand (which is made of rock fragments) and the piece of rock are gradually worn away. See how easy it would be to carry away the scraped material by putting it in a jar of water and swirling the water around. In a river this scoured material would be washed away.

Try various rocks such as sandstone and slate (from roofing tiles).

Block of wood with sandpaper wrapped around it.

Curves of stone

As rivers cut their canyons, they leave behind evidence of their former courses over millions of years.

Some rivers follow curving courses. Each curve, or **meander**, acts as a natural brake on the water as it flows seaward. A river flowing along a steep course will develop huge, sweeping meanders with loops that help to slow it down.

Goosenecks

A gooseneck is the name used when a river makes such a large curve that it nearly cuts back on itself. The San Juan River Canyon in Utah, shown in this picture, has cut some of the world's most spectacular goosenecks.

These goosenecks were formed as the river cut swiftly into the land millions of years ago. The curves then became fixed in position. Today the meander walls are 1000 ft high, and the river is still cutting down.

Why curves get bigger

As water flows down a river it is swirled to the outside of each curve. Here the water flows faster and so cuts into the bank eventually forming a steep cliff. On the other side of the curve where the water flows much more slowly there is very little cutting action taking place.

(In time some of the curves may cut through and this could give features like the natural bridges shown on the next page.)

For more information on meanders see the book River *in the Landshapes set.*

13

Natural bridges

Natural bridges are some of nature's most unusual landforms. They are produced only in places where there are thick layers of strong rocks such as sandstone and where rivers have cut goosenecks. Over thousands of years the river curves begin to meet, creating the thin rock walls in which natural bridges form.

How bridges grow and then fall
When the river breaks through a gooseneck it takes the shorter course which is under the new bridge, abandoning the old looping meander. As the river continues to cut into its new bed the hole gets bigger and the top of the hole eventually becomes a bridge.

A natural bridge is temporary. In time its surfaces crack and blocks fall away. Eventually it becomes so weak that it collapses.

The natural bridge in the photograph below and on the opposite page is in Natural Bridges National Monument, Utah.

Models like this one can be made easily using colored modelling clay. Simply roll the clay into thin sheets, then lay one sheet on another to show the different layers of rock. Use a blunt knife to 'erode' the 'land' and produce the landshape you want. The model is smoothed out using fingers.

Natural bridge

Looking through the bridge from below. Notice that the walls below the bridge are vertical. This shows that the river cut into its bed very quickly after forming the bridge.

The bed of the river is normally dry, but the pebbles and stones are readily picked up during floods and used to scour the sandstone rock bed and walls.

Natural arches

These extraordinary landshapes, formed high in the shared wall between two canyons, are far above where rivers can have any effect.

Most arches are formed in sandstone, a very tough rock made of tiny grains which can be loosened, or **etched**, from the cliff one at a time.

Where arches form

An arch often forms where beds of hard sandstone and soft clayey shale rest on each other. This means that arches can form at many levels, almost making a series of natural 'windows'. This double arch in Arches National Park is among the world's most spectacular high-level windows.

Grains are falling away, leaving surface pits.

This picture is of an arch that still has a thick layer of rock above. The flat ledge where the person is standing marks the junction between two different types of rock.

How arches form

Where **porous** sandstone rock rests on a bed of watertight rock like shale, rainwater will seep through the sandstone and then weep out where it reaches the shale.

Weeping sandstone dissolves the **cement** holding the sand grains and a cave begins to form.

Two caves forming on either side of a canyon wall may well break through to leave an arch. When the arch finally becomes too fragile the center collapses. However, the shape of the arch makes it strong. Arches can look very fragile but still remain standing!

Landscape Arch

The world's longest arch is in Arches National Park. It stretches over 291 feet, which is wider than 17 automobiles parked end to end.

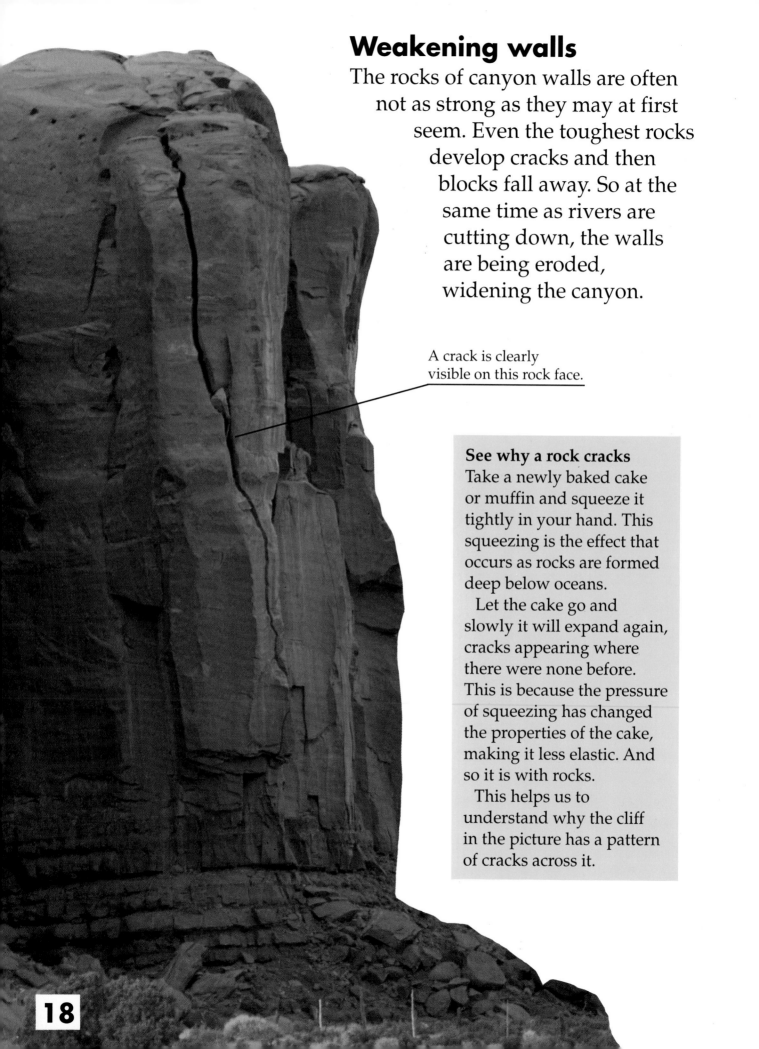

Weakening walls

The rocks of canyon walls are often not as strong as they may at first seem. Even the toughest rocks develop cracks and then blocks fall away. So at the same time as rivers are cutting down, the walls are being eroded, widening the canyon.

A crack is clearly visible on this rock face.

See why a rock cracks
Take a newly baked cake or muffin and squeeze it tightly in your hand. This squeezing is the effect that occurs as rocks are formed deep below oceans.

Let the cake go and slowly it will expand again, cracks appearing where there were none before. This is because the pressure of squeezing has changed the properties of the cake, making it less elastic. And so it is with rocks.

This helps us to understand why the cliff in the picture has a pattern of cracks across it.

The importance of cracks

If water can get into the cracks within a rock, the breakdown of the rocks is more rapid because when rocks are wet they crumble away more quickly. As the cracks widen, the rocks are held less firmly and eventually blocks can fall from the canyon walls.

Here you can see how water has etched the top of the rock, dissolving the cement between grains and forming a natural pond.

See where weeping rocks occur

Make a cliff from two sponges and a sheet of paper sandwiched between them. In a real cliff the sponges might be sandstone rock and the paper would be a clay rock called shale.

Trickle water onto the top sponge as though it were raining. The paper sheet (shale) will make the water 'weep' from the bottom of the upper sponge (sandstone) and spill down the cliff. This weeping water is one kind of **spring**.

What happens if you make some cuts in the paper and pour the water again? This would be the same as cracks forming in the shale rock.

Rainfall

Weeping rocks

River supply

Tray

Spectacular slopes

When you look through the pages of this book you will see many examples of steep canyon walls and cliffs. Rocks, weather and rivers all combine to cause these special landshapes.

Where there are layers of tough and weak rocks one above the other, the cliffs and walls are formed into a kind of natural staircase.

The flatter areas, or 'treads' of the staircase, are formed by the weaker rocks, like shales, which, once they have started to crumble, cannot stand up at a steep angle.

The cliffs, or 'risers' of the staircase, are made of much tougher rocks such as limestone or sandstone whose blocks fit together more like building blocks and they remain standing upright.

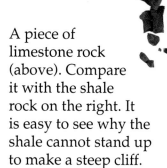

A piece of limestone rock (above). Compare it with the shale rock on the right. It is easy to see why the shale cannot stand up to make a steep cliff.

Make a moving staircase

To see how the staircase forms, you need some dry sand and some building blocks. The idea is to make a layer of sand and a layer of building blocks, all stacked inside a tube of cardboard.

When you slowly pull the cardboard upwards the lowest layers will 'erode' first, with the upper layers 'eroding' later.

Look carefully at the shape that forms and compare it with the natural landscape shown on these pages. What happens when you move some of the sand from the lowest layer?

Try using different thicknesses of each kind of rock. Does this change the shape of the cliff? Can you see examples in the pictures of the Grand Canyon on pages 32-35?

Can you improve on the design of the model to make it more realistic?

This sandstone cliff stands upright like a wall because it is made of a single layer of massive sandstone. There are few cracks for the weather to exploit.

This picture above shows broken limestone rock resting on the surface of a tread of shale. Most of the limestone has collapsed from the riser above.

Thin sandstone band

This part of the cliff is like a staircase because it is made of layers of shale and sandstone.

Shale tread

21

Chapter 2
Tracing nature's erosion cycle

Landshapes of change

As rivers cut into the land so tablelands are eroded away. First the land is deeply trenched to make canyons. Then the canyon walls widen so that the upland is separated into large blocks or mesas. Eventually the land will become mostly wide **plains** with few pieces of tableland.

This change from highland to lowland is called the **erosion cycle.** It may take hundreds of millions of years to complete.

Rivers are still cutting canyons into the sides of mesas and eventually they will cut them into many smaller blocks.

These rock towers shown above are pieces of the mesa that have been separated by river erosion. The bigger shapes are called buttes (see page 26), the smaller one to the right is a rock pillar (see page 28).

This is where the cliffs are most under attack (see page 24).

This mesa is a large flat tableland that extends for hundreds of square miles. It is encircled with steep, and often vertical, sides.

This place may once have been the site of a rock arch (see page 16).

These valleys will eventually widen to form a flat plain like that in the picture on the opposite page.

The canyon widens

The sheer walls of some dark and deep canyons show just how fast a river can trench into the landscape. But sheer walls will not last forever and slowly the rocks will crumble and collapse, helping the canyon to widen out. These make up the early stages in the erosion cycle.

The river that cut a canyon in this land disappeared. The canyon walls have crumbled away and left a long staircase of rocks.

This rock tower, or butte, is all that is left of the topmost layer of rock (see page 26).

The upper part of the slope is made of a single bed of sandstone which is very tough. This makes the vertical upper part of the slope.

The lower part of the slope is mostly made from shale rocks with only thin sandstone bands. That is why the slope is, on average, quite gentle.

The deep channels are the result of rivers cutting deeply into the rocks. The upper layers of rock have already been stripped away and the rivers are now eroding the newly exposed layers.

The rivers that flow here last for just a few days after each storm. In this picture only their dry beds can be seen.

25

Buttes: natural towers

As mesas wear away, the land changes from tablelands to isolated towers or buttes. These indicate the middle stages of the erosion cycle.

The towers also give an indication of how far the tablelands used to extend before they were carved out by rivers.

Leftovers
This landscape shows how buttes are often the last reminders of a former tableland. All around these buttes in Arizona's Monument Valley is flat land that the rivers have made. In this way nature replaces one area of flat land with another, at a lower level.

How buttes form

Buttes form when the shared wall of two canyons is worn away. They are the land version of sea stacks that you sometimes see along cliffed coastlines.

Buttes keep their cliff-like faces when there is a thick tough band of rock on top of a weak one.

The weak rock is carried away during rainstorms, undermining the tough rock which then falls away in slabs.

Pillars and needle rocks

Many needle rocks form as the last remnants of buttes. These show the last stages of the erosion cycle.

Pillars and needle rocks stand up in the same way as a chimney made of building blocks. Losing rock from the outside does not weaken the structure that is left.

This aerial view of Arches National Park shows balancing rocks and arches forming as the sandstone is almost completely eroded away.

Make a balancing rock
You can make a balancing rock using pieces of modelling clay. First make the modelling clay into blocks, then make a large ball and slightly flatten one side. This can then be added to the pile of blocks. When you have finished the tower should look like the picture on the right.

Try to get the large block to sit on top of the others just like the ones shown here. Your model will then let you understand how the rock balances without falling down.

These rock pillars show where the butte used to be. Erosion has taken away the rock that used to stand between the mesa and the pillar.

This is the edge of a mesa.

Monument Valley, Arizona

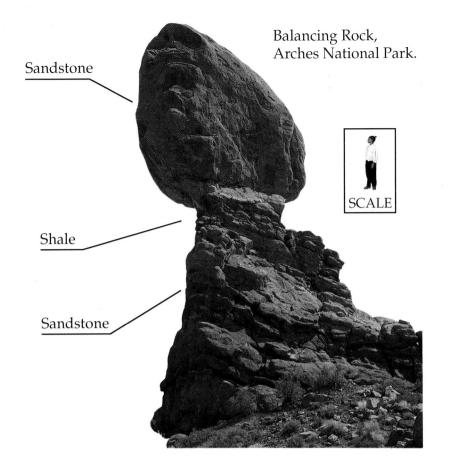

Balancing Rock,
Arches National Park.

Sandstone

Shale

Sandstone

SCALE

This is what
the sandstone
rock looks like.

These very thin rock pillars
are called needle rocks. They
still stand because, like a
chimney, the rocks from
which they are made are
strongly cemented together.

29

Chapter 3
Canyonlands of the world

Canyonlands National Park

Canyonlands is an area of outstanding natural beauty that has been preserved as a national park in the State of Utah, USA. For over a million years the land has been slowly rising, forcing the Green River and many of its side rivers (**tributaries**) to cut deep canyons in the land.

The Green River feeds, in turn, into the Colorado and its waters eventually flow through the Grand Canyon (see pages 32-35). Canyonlands gives a glimpse of what the Grand Canyon may have looked like before it cut the great valley that we see today.

Small rivers do not have the energy to cut quickly into a tableland and many are still in their 'youthful' stage. But notice how the **box canyon** has worn back and widened to match the way the main river has cut down.

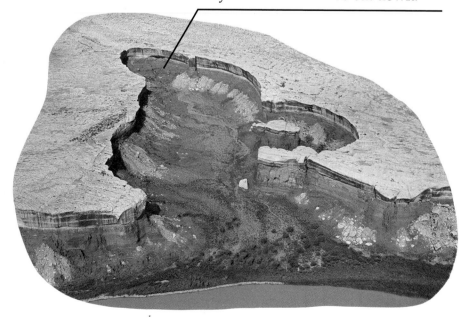

The canyon widens as the rock walls crumble and the waste is carried away by the river.

The rocks are red because they contain iron stain.

This is where box canyons form. You can see a box canyon more clearly in the picture on the facing page.

The Green River
Here the river has cut deeply into the land. The toughest rocks still stand as cliffs, while the weaker have long since begun to crumble away.

The climate is too dry to form soil and too dry for many plants to grow, so the strength of each rock shows clearly, and you can count the number of rock bands.

The Grand Canyon

The Grand Canyon is the world's biggest canyon, cutting a huge gash in the tablelands of northern Arizona, USA.

The canyon has cut through the rocks of this high tableland just like a knife cutting through a sandwich. Many rock types make up this canyon and they are clearly seen, layer upon layer. White limestones, that once formed as **reefs** below the sea, now make the topmost layer. Below them lie deep red sandstones, reminders of a former desert. Mixed up with the limestones and sandstones are shales, crumbly rocks made from ancient ocean bed muds. They are the weakest rocks and they crumble readily.

The oldest – and deepest – rocks of the Grand Canyon are made of the remains of ancient volcanos. Now, as the Colorado scours ever deeper into the floor of the canyon, they are being exposed for the first time in a billion years.

Kaibab Limestone (white)

Toroweap Sandstones (tan)

Coconino Sandstone (cream)

Hermit Shale (bright red)

Supai Silts and Shales (red)

Redwall Limestone (red)

Bright Angel Shale (grey)

Tepeats Limestone and Vishnu Schist

The upper rim of the canyon regularly breaks away and its shattered rocks tumble to the canyon floor below. Here a massive crack in the limestone (just below where the person is standing) shows where a collapse will soon take place.

This is the South Rim, the picture shows a view to the west.

Inside the Grand Canyon

The Grand canyon is so big that it takes a whole day to walk from the rim to the river and back again. But as you enter the canyon on one of the trails, and the rim rears up like a huge wall above, you begin to see much more detail of each rocky ledge and slope until finally the great river itself comes in to view.

One way to ease the burden of a long walk is to use mules.

Here a tributary box canyon is eating back towards the South Rim. See how a tough band of rock is slowing its progress. During storms the edge to the box canyon will form the lip of a waterfall.

The Colorado River is a tan color because it carries so much rock and soil scoured from its bed and banks. Even stones and dust dislodged near the rim by people and mules eventually reaches the river and are carried to the Pacific Ocean, hundreds of miles away.

The white Kaibab Limestone rock is on the North Rim, 13 ml away.

This rocky ridge is the result of erosion of two neighboring box canyons.

New words

box canyon
the name given to a small canyon that is tributary to a large canyon. Box canyons end in steep-sided cliffs

butte
a natural tower of considerable size and which is no longer part of the main tableland. There is no special size for a butte. It changes to a pillar rock when the butte is nearly worn away

cement
many rocks are stuck together with natural cements as they are being formed deep under the sea. One of the most common cements is natural lime. When a rock with a weak cement is exposed to the weather, the natural lime cement dissolves and the sand grains fall away. This is an important reason for the growth of arches

erode
the process of wearing the land away. There are two parts to erosion: first the rock must be loosened, and then it must be carried away. A river erodes its banks because it loosens the material and carries it away instantly. However on a slope frost may loosen a rock but it may be many years before it falls away

erosion cycle
the sequence of events that occurs when rivers, landslides, rock falls and other processes erode an area of high land, gradually reducing it to a plain

etch
a word used to describe erosion which picks out the tougher and weaker rocks clearly

meander
a natural curve or bend that a river makes as water flows in a channel. If the curves swing almost completely back on themselves, the meander is called a gooseneck

mesa
the edge of a tableland that has been eaten into by many canyons

plain
a region of low lying land that has been produced by river action. Plains are not absolutely flat, but have many gentle slopes on their surfaces

porous
a name used to describe a rock that has many small gaps between its rock particles and which can therefore soak up water

reef
a large bank of limestone produced by the skeletons of corals and other animals in a tropical sea

spring
a place where water seeps from a rock. If a spring occurs high up on a cliff, the cliff is said to weep

tableland
a name for a large area of high level land. Another name for such a land is a plateau

trench
a gorge cut into the land by river action

tributary
small streams or valleys that cut into a landscape and help deliver surplus water to the main river or valley

Index